DRAGONS

Myths, Legends & History

KIV Books

Copyright © 2019

Copyright © 2019 KIV Books

All rights reserved. This book or any portion thereof may not be reproduced or used in any manner whatsoever without the express written permission of the publisher except for the use of brief quotations in a book review.

Disclaimer

This book is designed to provide condensed information. It is not intended to reprint all the information that is otherwise available, but instead to complement, amplify and supplement other texts. You are urged to read all the available material, learn as much as possible and tailor the information to your individual needs.

Every effort has been made to make this book as complete and as accurate as possible. However, there may be mistakes, both typographical and in content. Therefore, this text should be used only as a general guide and not as the ultimate source of information. The purpose of this book is to educate.

The author or the publisher shall have neither liability nor responsibility to any person or entity regarding any loss or damage caused, or alleged to have been caused, directly or indirectly, by the information contained in this book.

Table of Contents

INTRODUCTION 5

WHAT ARE DRAGONS? 7

WESTERN DRAGONS (EUROPE AND THE AMERICAS) 11

EASTERN DRAGONS 19

PROBABLE ORIGINS OF DRAGONS 33

DRAGONS IN POPULAR MEDIA 39

DRAGON "SIGHTINGS" IN MODERN TIMES 49

Introduction

What is it about dragons that almost every country in the world has their own version of this mythical creature? Why is it that countries that are on opposite sides of the world have dragons that have relatively the same features? Is this evidence that these mythical beasts did not just exist in the imagination of people from centuries past?

All these questions and more will be answered in this book. If you have always been fascinated with dragons, and you want to learn more about them, then this is the book for you.

Here, you will learn about the different kinds of dragons, by type and according to the region where they originated. You will learn about the differences between the malevolent European dragons and the (mostly) benevolent dragons of the East. You will also learn about the different dragons that have been featured in popular culture, from books, movies, and even video games.

Dragons have been mainstays in past and present culture, and it seems that there are no signs of their popularity waning any time soon.

What are Dragons?

"If you want to conquer the world, you best have dragons."

- George R.R. Martin – A Dance with Dragons

What are Dragons?

Dragons are one of the most popular mythological creatures that are still reviled to this day. Almost everyone has heard of dragons. When people imagine what a dragon looks like, they would almost always envision a huge, winged, reptilian creature that not only flies but also breathes fire. Most people think that dragons have a notorious habit of raiding castles, kidnapping princesses, and sleeping on top of a huge pile of treasure. However, like all mythological creatures, the legendary dragon did not start out that way.

Almost as long as there have been people on the earth, so does legends and myths about dragons. For instance, many Early Mesopotamian cultures have ancient stories that have been passed down for hundreds of generations, many of them talk about the mighty storm gods saving the populace from evil giant serpents. These "serpents" are quite scary. Some accounts tell of dragons that have iridescent scales, and some have the uncanny abilities of flight and fire breath, which are most likely the basis for the modern interpretation of dragons.

What do dragons actually look like? There is no one definitive description of dragons, but there are some universally accepted features that these mythological creatures are said to have. For instance, many cultures believe that dragons come in the form of giant serpents, and that they are particularly fearsome in nature. This is also the reason why they are called such, since the word "dragon" was said to be derived from the Greek word "drakon", which means "large serpent".

Most dragons are described to be evil, especially if you are talking about the ones that came from Western cultures. However, in Asia, in China most particularly, dragons are said to be wise and benevolent creatures.

Etymology of Dragons

The word dragon is derived from the ancient Greek word δράκων (pronounced drakōn), which means "huge serpent" like a python for instance.

Dragoons, the armored infantry division that moved around on horseback and fight like foot soldiers, got their name from the early form of musket (a wide-bore firearm) that these units used, which was called "dragons" because they "spat fire" and produced a loud noise when fired.

Earliest Known Information About Dragons

Some of the earliest written records that have any mention of western dragons came from the ancient Greeks. Herodotus, the *father of history*, wrote about his trip to Judea in circa 450 B.C., and in his records he mentioned that he heard tales from the locals about small, reptile-like creatures that can fly, which they called dragons. Herodotus also wrote about the time he saw large bones that people said came from an ancient dragon.

Herodotus was not the only Greek to talk about dragons. In Greek mythology, there are many stories about great snakes or dragons, most of the tales tell about these wicked creatures guarding a huge hoard of treasure.

Legends also say that the first Pelasgian kings of Athens were hybrid humans that are half man and half snake. Another legend tells about how the Greek hero Cadmus, upon the instructions of the goddess Athena, slain the water dragon that guarded the Castalian Spring. Athena instructed Cadmus to sow the teeth of the large serpent in the ground, and from those teeth sprouted a race of warrior men, the Spartes, who then assisted him in the construction of the citadel of Thebes, and they became the founders of the noble families of the great city.

These are just some of the ancient Greek records that have mentioned dragons. There would be a more detailed discussion about this later on in this book

During the middle ages, in the medieval times, dragons were used as symbols of treachery, envy, and anger. Some were also used to symbolize great calamities. Some used dragons to symbolize self-indulgence and oppression, the medieval Christian churches also used dragons as a symbol for heresy.

However, not all medieval dragon symbolism were not all evil. Some dragons symbolized independence, strength, leadership, and wisdom. Tales of heroes who have slain dragons not only gained access to the riches that the creatures have guarded for hundreds of years, it also meant that said heroes managed to best the most cunning and wise of all creatures.

Myths Associated with Dragons

Although countless stories tell of dragons that guard enormous piles of treasures, there are also quite a lot of myths that surround the creature itself. For instance, there are many myths about dragon's blood. One myth states that when you dip a sword or a dagger in dragon's blood, the wound made by the weapon will never heal. There is also a myth that when a person drinks the blood of a dragon, he or she will gain the ability of clairvoyance, in other words, they will be able to see into the future.

In Eastern cultures, it is thought that dragons have the magical ability to shapeshift. There are many tales from Asian countries that tell about dragons that could change into a human form so that they can live amongst them.

Western Dragons (Europe and the Americas)

"What is better? To be born good, or to overcome your evil nature through great effort?"

- Paarthunax – The Elder Scrolls V - Skyrim

All over Europe, you will hear a lot of different folk tales and songs that involve dragons in one form or another. You will also find that different European cultures have different takes on dragons and what they do, but there are also some similarities that makes it seem that the reality of dragons is possible.

Here are some of the different nuances between the different versions of dragons from the different countries in Europe:

Slavic Dragons

The most common dragons of Slavic mythology are the *zmeys* (Russian), *smok* (Belarussian), and *zmiy* (Ukrainian). These dragons are usually seen as protectors of crops and as a symbol of fertility. From the get go, these European dragons are already quite different from their counterparts as they are benevolent rather than malevolent.

These dragons are almost always portrayed as having three heads, and a conglomeration of different animals, usually a combination of snakes, humans, and birds. These dragons do not usually stick to one form and would frequently shapeshift. These dragons are generally male, and they are quite sexually promiscuous; there are stories wherein they would mate with humans. The zmeys, smok, and zmiy, are associated with the elements of fire and water, both important for the survival of humans, especially in the tough Slavic environment.

Sometimes, there appears a Slavic dragon that is portrayed as an evil, four-legged monster that have very few, if any, redeeming qualities. These dragons are somewhat smart, but not genius-level smart. They would usually take entire villages and towns hostage and would require tributes, like maidens in exchange for food and sometimes gold. The number of heads of these types of dragons vary from three to seven, sometimes even more. These dragons are very difficult to slay as the heads would grow back after they are chopped off; the only way to prevent the heads from coming back is to cauterize the wound immediately using an intense flame (a treatment that is also used to kill the hydra of Greek mythology). The blood of these dragons are also said to be so poisonous that it would not seep into the ground as the earth itself is reviled by it.

Another kind of European dragon is the Polish Wawel Dragon, also known as *smok wawelski*. The Wawel dragon supposedly terrorized ancient Krakow and would retreat back to its home in the caves lining the Vistula river bank, somewhere under the Wawel castle, hence its name.

According to Polish lore that is based on the Book of Daniel, the Wawel dragon was killed by a boy with a sheepskin full of sulfur and tar. The dragon got so thirsty after eating the sheepskin that it could not stop drinking, until it literally burst at the seams from drinking too much water.

If you want to see what the Wawel dragon looked like, you should go visit Krakow, there is a stylized metal sculpture there of the legendary dragon that is favorite of tourists and locals alike. The neat thing about this sculpture is, like the "real" Wawel dragon, this one breathes fire every couple of minutes or so.

The Wawel dragon is not the only dragon in Polish culture; they have the basilisk that is said to live inside the cellars of Warsaw, and the Snake King from local folk tales.

Germanic and Norse Dragons

In the Germanic and Norse cultures, dragons are known as "Lindworm", which is a variant of serpent-like mythical creatures known as wyverns. Lindworms are usually depicted as gigantic, monstrous serpents that have legs and bat-like wings. These dragons look more like winged snakes than the traditional European dragons.

Lindworms were believed to be malevolent, and they would usually steal cattle and other livestock. People believed that seeing one in person is a bad omen. Lindworms are somewhat greedy creatures, and they are often believed to guard huge stashes of treasure, and they usually live in caves deep underground.

Lindworms, according to Germanic and Norse folklore, were once people who are so greedy that they transformed into creatures that best signify their sins. Some of the most popular legends concerning lindworms include the one about Jormungand, the love child of Loki and a giantess. When Odin found out about it, he threw it into the sea surrounding Midgard, where it ate so much fish that it grew to such a size that it can fully encircle the entire earth. Another popular legend is that of Fafnir, a human man who killed his own father just so he could inherit his vast riches. Because of his greed, and overwhelming desire to protect his ill-gotten wealth from others, he transformed into a lindworm.

British Dragons

There have always been dragons in the lore of the British. However, the dragons of British legends were more like the wyverns that are in the tales from Central Europe. Just like the wyverns, British dragons could also fly and breathe fire. Perhaps the most famous dragon in all of England is the one slain by the country's patron saint, St. George.

These days, there are two dragons that are in Great Britain; there's the White Dragon, which is symbolic of England, and the Red Dragon, which appears on the Welsh flag. There is an ancient legend in Britain about a red and white dragon fighting to the death, after the bloody, long-drawn battle, the Red Dragon came out victorious. The Red Dragon is symbolic of the Britons, who are today represented by the Welsh, and the White Dragon represents the Saxons, and now the English (they were the invaders of the Southern British Islands).

Basque Dragons

There are not that many dragons in Basque legends, however thanks to writers such as Juan Delmas and Chao's interests in them, they managed to preserve the legend of the *Herensuge*, which means the "third" or "last serpent". The Herensuge, is an evil spirit that took the shape of a giant serpent-like creature that terrorized towns and villages, killed countless numbers of livestock, and trick and mislead the people.

The best known legend about the Herensuge must be the one where St. Michael descends from heaven to fight against it, but the battle almost ended in a deadlock. It was only until God helped St. Michael did the archangel defeat the evil dragon.

Italian Dragons

The legend of St. George is well-known in Italy, but there also were other saints who fought dragons. For instance, Saint Mercurialis, the first bishop of Forli, fought and killed a dragon to save the town. Another example is Saint Theodore of Tyro, the first patron saint of the city of Venice, was a well-known dragon slayer, and he has a statue of him slaying a dragon on top of one of the pillars in Saint Mark's square.

Dragons is Christianity

There are no direct references of dragons in the Bible, but there are some creatures whose descriptions seem to fit what dragons traditionally look and act like.

For instance, the monster Leviathan that swallowed poor old Job resembles what a dragon looks like. There is also one mentioned in the book of Revelations, which is a gigantic red beast with seven heads, and has a tail that is so long that it sweeps away a third of the stars from the sky and send them crashing onto the earth. In many translations of the Bible, the beast in Revelations is described as a dragon.

The church during the Middle Ages interpreted the Devil as being in cahoots with the serpent who tempted Eve and caused her and Adam to be cast out of the Garden of Eden. The demonic foes of God, or good Christians, are commonly portrayed as having dragon-like qualities.

The traditional image of a dragon, a vile creature whose greed knows no bound, lives underground and guards a huge horde of treasure, is also regarded as the embodiment of sin; the sin of avarice to be exact. It was also around the Middle Ages that Catholic literature and iconography started depicting some saints as dragon slayers.

In addition, some Christian authors mentioned that dragons were good once, that is before they were cast away from the Garden of Eden right after God exiled Adam and Eve. Another argument that these Christian authors had for their claims that there were once good dragons is the fact that they were one of the creations of God, and actually resided in the Garden.

Dragons in Heraldry

You can often find images of dragon, or dragon-like creatures in heraldry all over Europe, most notably in

Great Britain and in Germany. The most common creature in heraldry are wyverns, which are dragons that have two rear legs and two wings. Wyverns are symbols of strength and protection, but many of them also symbolize revenge and vengeance. The traditional four-legged, two-winged dragons are the second most popular, and they symbolize wealth and power.

In Great Britain, images of dragons were first made famous by Uther Pendragon, King Arthur's father, who had a dragon on his family crest. Also, the legend of Saint George and the dragon is also one of the most famous folk story in Britain. Although dragons used in heraldry do symbolize positive characteristics, many cultures in Europe still view dragons as malevolent creatures.

Eastern Dragons

"Fairy tales are more than true: not because they tell us that dragons exist, but because they tell us that dragons can be beaten."

- Neil Gaiman, Coraline

Europe and the entire western hemisphere does not have a monopoly on dragons, not by a longshot. There are literally hundreds of different types of Eastern Dragons, and you can find at least one semblance of them in every country in Asia and other regions in the East.

Chinese Dragons

The Chinese Dragon (spelled phonetically as *Loong, Long, or Lung)*, also called Oriental or Eastern dragon, is a mythical creature that appears in many East Asian cultures.

The Chinese dragon, unlike its Western counterpart, is not necessarily evil. These dragons are usually powerful spiritual symbols that represent the changes in seasons and supernatural forces.

You can easily distinguish the Chinese dragons from Western dragons, thanks to their long and serpentine bodies that are usually wingless, and they also have anthromorphic faces with beards.

In some Asian cultures, dragons play an important part in the mythology of creation. In general, eastern dragons are benevolent and very powerful, and the bringer of good fortune. This is why the Chinese dragon was adopted by ancient emperors as a sacred symbol of heavenly power.

Descriptions

The earliest images of dragons in Oriental cultures come in the form of totems, which are stylized images of different animals. One of the earliest forms of totem dragons are the "pig dragons", which are creatures with coiled and elongated bodies with heads resembling that of a boar's. The early Chinese character for "dragon" also shared the same coiled form as that of "pig dragons", these images also appeared in ancient jade amulets that came from the Shang Dynasty. In the early days, dragons were depicted as conglomerations of different kinds of animals, and over time, the image of the traditional dragon has evolved into a single, unified visage of a mythical creature.

In general, the appearance of Eastern dragons can be seen as a combination of different animal parts: the serpentine body of a snake, the thick scales of a carp, the tail of a whale, antlers of a stag, the sharp talons of an eagle, and the eyes of a lobster. Sometimes, dragons are also depicted holding a flaming pearl under its chin, and some Chinese dragons have bat-like wings growing out of their front legs.

Although Chinese dragons are typically shown as wingless, they are almost always depicted while in flight. Dragons are also said to have an almost unlimited variety of supernatural powers; some accounts detail how dragons can take the shape of a small silkworm, and some tales say that they can expand to the size of the entire universe; some tales tell of how dragons can fly among the clouds, turn into water or fire, can go invisible, and some can even glow in the darkness of night.

Origins

There is no clear consensus as to where the legend of the Chinese dragon began. Just like the origins of the European dragons, the Chinese dragons are said to be inspired by the discovery of dinosaur fossils made by the ancient Chinese. When they saw the gigantic bones of incredible looking creatures, it fueled the belief that there once existed giant reptilian creatures that have supernatural powers. There is also a chance that Marco Polo, a famed European explorer, might have been the originator of dragons in Europe when he supposedly saw a dragon during his travels in the East.

As mentioned earlier, many scholars suggest that the Chinese dragon has its roots from the ancient totems used by the indigenous tribes in China. Scholars claim that Huang Di, the legendary fist emperor of China used a snake in his coat of arms, and whenever he defeated an enemy tribe, he would incorporate their emblems into his own, which might explain why the Chinese dragon somewhat looks like a mash-up of different animals stuck on the body of a snake.

Some scholars also suggest that the image of the Chinese dragon came from the depictions made by artists using only eyewitness accounts as their basis. For instance, there is the legend of Zhou Chu, a legendary hero who purportedly killed a dragon that infested the waters near his village. In this case, the "dragon" seems to be a crocodile.

The famous Chinese archaeologist Zhou Chongfa theorized that the inspiration for dragons was lightning, mainly because the way the Chinese pronounce dragon, which is "long", somewhat resembles the sound of thunder. This is why generations of farmers believed that the dragon represents the coming of the rain. Zhou claims that the rise of farming and animal husbandry in lieu of hunting and gathering as a means to acquire food, caused the people to pray for good weather so that their crops and livestock can thrive, and thus formed the image of the all-powerful and benevolent dragon.

Chinese Mythology and Cultures

Unlike the way that Europeans depict dragons, whereas they are mostly evil, many versions of the Oriental dragons are actually powerful spiritual symbols, and that they are mostly good and benevolent.

The Chinese revere dragons in many different ways. For instance, tigers are seen as the eternal rival of dragons, which is why you will see countless numbers of artworks depicting these two creatures in epic battles. In fact, in the Chinese martial arts, the Dragon Style is a method of fighting that is based upon the understanding of movement, while the Tiger Style is a method based on brute strength and mastery of a technique.

Chinese dragons in popular belief have a strong association with water. Dragons are believed to rule over moving bodies of water, like rivers, streams, seas, and waterfalls. They can also take the form of water spouts (tornadoes that appear over water). Dragons are also rulers of weather, and in this capacity, they are more anthromorphic in form, and they are often shown in the form of a human, often dressed in a king's raiments, but with the head of a dragon.

The Dragon Kings

In Chinese culture, there are four major Dragons, called Dragon Kings, each of them represents the four seas surrounding China: the East Sea (the East China Sea), the West Sea (the Indian Ocean and beyond), the South Sea (the South China Sea), and the North Sea (sometimes seen as Lake Baikal). Because of this, the Dragon Kings are in charge of all water-related weather phenomenon.

In the ancient times, many villages in China, especially those living along the banks of rivers, had temples that are dedicated to their local Dragon King. Whenever there was droughts or flooding, the local government would conduct religious rites and place offerings to appease the Dragon King, or to ask for a blessing.

In Chinese consider the number "9" as lucky, and they are connected to a certain extent with Chinese dragons. For instance, Chinese dragons often has nine attributes, and typically has 117 scales; 81 (9x9) for male, and 36 (9x4) for female. This is also why there are nine Chinese dragon forms, and why the Dragon King has nine children. There is a screen wall, called the Nine Dragon Wall, with images of nine dragons in the imperial palaces and gardens. And because the number 9 is considered a number fit only for the emperor, only the most senior of court officials are allowed to wear robes that have nine dragons, and their robes have to be completely covered by a surcoat. Lower ranking officials can only have eight or fewer dragons, and also have them covered with surcoats. Even the emperor himself can only wear his dragon coat with at least one of the dragons covered up with another piece of clothing.

The Nine Type of Chinese Dragons

Tianlong – The Celestial Dragon

Shenlong – The Spiritual Dragon

Fucanglong – The Dragon of Hidden Treasures

Dilong – The Underground Dragon

Yinglong – The Winged Dragon

Jiaolong – The Horned Dragon

Panlong – The Coiled Dragon

Huanglong – The River Dragon

The Dragon King

The first four dragons are the most popular in Chinese culture. The celestial dragons are the ones that protect the gods' palaces in heaven, the spiritual dragons has command of the winds and the rain, the dragons of underground treasure protect wealth, and the underground dragon controls the geological events, like earthquakes and landslides.

Aside from the Nine Dragon Types, there are also Nine Dragon Children that you can usually use in old Chinese buildings and structures:

Bixi, the first son, looks like a gigantic tortoise and can carry a lot of weight. This is often found carved into the stone bases of monuments.

Chiwen, the second son, looks like a wild animal and can see very far. This is often seen as gargoyles placed on the roofs of palaces.

Pulao, the third son, looks like a small dragon and he likes to roar. This is why you can find symbols of it carved into temple bells.

Bi'an, the fourth son, looks like a tiger, and he is very powerful. You can find images of this dragon son painted on prison doors to prevent the prisoners from attempting to escape.

Taotie, the fifth son, loves to eat and you can find his image printed on tableware, cookware, and silverware.

Baxia, the sixth son, loves the water, which is why his image can be found on bridges and canals.

Yazi, the seventh son, loves to kill, which is why he is found on many bladed weapons.

Suanni, the eighth son, resembles a lion and likes to smoke; he also likes fireworks. His image can usually be found in most incense burners.

Jiaotu, the ninth son, looks like a clam, and he does not like anyone to disturb his sleep. This is why you will usually find his image on front doors or right on the doorstop.

The Chinese Zodiac

The dragon also has an important role as one of the animals in the Chinese zodiac, which is used to designate the years in the Chinese Lunar Calendar. The dragon is the fourth sign of the zodiac (because he was the fourth to get to the palace of the Jade emperor when he summoned all the animals on earth), and represents the years 1902, 1914, 1926, 1940, 1952, 1964, 1976, 1988, 2000, and so on. The year of the dragon is considered as one of the ideal signs to be born under, because it represents ambition, being headstrong, and courageous. However, those born under the sign of the dragon can also be quite reckless and stubborn.

The dragon, notably the Azure Dragon, Qing Long, is one of the four primary Celestial guardians; the other three being Zhu Que (Red Phoenix), Bai Hu (White Tiger), and Xuan Wu (Black Tortoise). The Azure Dragon is associated with the Eastern direction and the wood element.

During auspicious occasions, like the Chinese Lunar New Year and the grand opening of shops and during house blessing ceremonies, the festivities would often include a special dragon dance. These are when life-size cloth and wood dragon puppets are brought to life by a team of trained dancers performing choreographed moves accompanied by loud drums and other traditional musical instruments.

During special festivals, like the Duan Wu festival, which usually happens right before the start of the summer solstice, people would hold dragon boat races on the major rivers. Dragon boats are usually manned by 12 rowers and a navigator and drum beater at the helm; the drummer helps keep the rowers in cadence with each other to make them row more efficiently.

Dragons as Symbols of Imperial Authority

In ancient China, dragons became symbols of the the royal imperial family. In fact, at the end of his reign, the first emperor, Huang Di, ascended to the heavens and transformed into the dragon as represented in his royal emblem. And because the Chinese believe that Huang Di is their common ancestor, the citizens usually refer to themselves as the "descendants of the dragon". This is also the reason why the Chinese use the image of the dragon as the symbol of the emperor.

The dragon, specifically the golden dragons, the ones with five talons on each foot, was the symbol of the Chinese emperor for a lot of the past dynasties. In fact, the imperial throne was referred to as the "Dragon Throne", and in the latter parts of the Qing Dynasty, the Chinese flag adopted the symbol of the dragon for its national flag. At the time, it was also a capital offense for commoners to wear clothing that has any imagery of dragons.

Other Asian Cultures' Dragons

The Chinese mythology is most probably the origin of the other dragon imagery in all of Asia, which is probably why the dragons from other countries are similar in appearance to the Chinese dragons.

Japanese Dragons

The Japanese dragon, locally called *ryu* or *tatsu* looks quite similar to the Chinese dragon in appearance; it also has a long serpentine body, it is wingless, has small clawed legs, has horns or antlers, and has mammalian facial features. The only obvious artistic difference is that Japanese dragons have three claws per hand, whereas the Chinese dragons have five claws. In Japan, just like in China, dragons were once connected to the rain, seasons, and water in general, which is why the Japanese also hold them in high regard.

Dragons are also an important part of ancient Japan. It was believed that Japan's first emperor was the child of Hoori, a hunter, and the daughter of dragon king, Ryujin. Ryujin was a mythical deity that lived in an ocean palace that was near Okinawa, and just like the other ancient dragons, he could control the seasons.

Another Japanese legend involves the dragon king and Otohime, his daughter. A fisherman name Urashima, while he was out fishing in the ocean, caught a sea turtle, which is an animal that the ancient Japanese consider as sacred. Urashima spared the poor creature and released it back into the ocean. The tortoise was actually Otohime, and to show her gratitude, she allowed Urashima to visit the Dragon King's underwater palace.

When Urashima visited the palace, he immediately fell in love with Otohime and the two got married. Eventually, Urashima asked Otohime if he could go back to the land so he can visit his family. Before he left the palace, Otohime gave Urashima a box that he needs to carry, but never open. Otohime did not mention the fact that once he leaves he could never go back to the palace.

When he returned to his home, he was surprised to find that more than three hundred years have passed by since he went to the Dragon King's palace. Distraught upon the loss of all his family and friends, and also because he does not know how he could go back to the Dragon King's palace and Otohime, Urashima opened the box his wife gave him, and out flowed all of the time that he managed to escape, causing him to age rapidly and die.

There are other dragoons that appear in Japanese mythos, like in the story titled, "My Lord Bag of Rice". The story tells of a hero that is tasked to kill a gigantic centipede that is eating the children of the Dragon King of Lake Biwa.

In addition, there are also some dragons that are considered as evil in Japanese folklore, like the Yamata-no-Orochi, an eight-headed and eight-tailed dragon that devours maidens, which was slain by the hero Susanoo.

Vietnamese Dragons

In Vietnam, dragons (*long* in Vietnamese), is one of the most sacred symbol and the most prevalent image that can be found in cultural architecture in the country. In fact, dragons are more common in Vietnam than in most other countries.

Just like in other Asian countries, Vietnamese dragons are believed to bring rains that are essential for agriculture, which is the main source of livelihood in the country. The dragon is also representative of the King, the country's power and prosperity, and it is the symbol for "yang", which represents the male.

In Vietnamese legends, the people of the country are the descendants of the children of the Dragon King *Lac Long Quan* and the fairy *Au Co*. From these 100 children born from the Dragon King was the king *Lac Viet*, the very first emperor of the dynasty of Vietnam.

Other Vietnamese legends portray dragons as the protector of the Vietnamese people. Legends tell that a dragon protected Vietnam from the hordes of Genghis Khan by decimating the enemy army by its fire breath, and by advising *Ly Thai To* to move Vietnam's capital away from the nearly overflowing waters of the Red River.

Korean Dragons

The Korean Dragon (*yong*), just like the Japanese dragons, are very similar to the Chinese dragon. Just like the Chinese dragons, Korean dragons are benevolent beings that are related to water, the weather, and agriculture. These dragons are said to reside in bodies of water like rivers, lakes, oceans, and also in the deep ponds found inside mountains.

Just like with the Chinese, Koreans also hold the number nine as sacred. Korean dragons are said to have 81 (9x9) scales on their backs. But, Korean dragons are always depicted as not having any wings at all, and they occasionally hold a "dragon orb" also known as the *Yeo-ui-ju* in one of its clawed hands. Legends say that whoever can get hold of the dragon orb, he or she will be blessed with near omnipotent power.

The Korean dragon is seen by the Korean people more as a spiritual symbol that lives inside the mystical palaces that are beyond the realms of people's minds. Later, with the additional influence of China, the dragon also were connected to the emperor.

Probable Origins of Dragons

"A man who has a large and imperial dragon grovelling before him may be excused if he feels somewhat uplifted."

- J.R.R. Tolkien – Farmer Giles of Ham

The dragon is the most popular creature in legends and in human mythos. Dragons are almost always depicted as a gold-hoarding serpent sleeping deep in the bowels of the earth, a fire-breathing gigantic monster, and other kinds of creatures. Even today, dragons still continue to fascinate people around the world. However, where did the idea of dragons originate? There are no definite time and place where dragons first came to be, however, tales concerning dragons have existed since the time of the ancient Greeks and Sumerians.

Over the years, researchers have come up with some explanations as to why people started believing in dragons. Here are some of the creatures that people (probably) mistook for dragons:

Crocodiles

The saltwater and the Nile crocodile are the largest known reptiles currently found in the earth. Saltwater crocodiles, despite the name, have a broad range of habitats, from the eastern regions of the Indian Ocean, all throughout Indonesia, and along the northern coastline of Australia. On the other hand, the Nile crocodile can be seen along riverbanks, lakes, and the deep marshes of the Sub-Saharan Africa.

A thousand or so years ago, both species of crocodile might have a wider habitat. Archaeological evidence shows that some Nile crocodiles lived as far as the northern part of the Mediterranean, which means that the ancient inhabitants of Italy and Greece could have been threatened, and they could have easily been mistaken for dragons.

In addition, Nile crocs can grow up to 20 feet in length, and they can lift their trunks off of the ground, which might be the reason why dragons are often depicted to be rearing up to attack the knights that have come to do battle with them. In addition, saltwater crocodiles are such powerful swimmers that they can propel themselves out of the water, which might be mistaken for dragons rising up from the water.

Dinosaurs

A lot of archaeologists believe that the early people might have mistaken dinosaur fossils for dragon bones. For instance, the Qijianglong, a dinosaur fossil that is believed to live 160 million years ago, and this creature measured almost 50 feet in length. This fossil was discovered by a couple of construction workers in China. One by one, the workers uncovered huge vertebrae set in a row buried in the ground. Today, people can actually tell that these fossils came from dinosaurs, but in the early days, with no prior scientific knowledge, it is easy to mistake the bones of dinosaurs as that of ancient dragons.

To further back up this theory, there are plenty of evidence showing that the ancient Chinese have been studying dinosaur fossils as far back as the 4th century BC.

Whales

Another theory for the origin of dragons is that early people found the skeleton of beached whales and they have no idea where it came from. Because whales spend a majority of their lives underwater in the high seas, and the ancient humans, who had no knowledge of the gigantic creatures that lived in the oceans, and they just assume that the gigantic bones washed up on the shore came from dragons.

In the Bible, in the Book of Job, the Leviathan (which was believed to be Cetus, a sea monster sent by Poseidon to attack Ethiopia) is pictured as a dragon:

12 "I will not fail to speak of Leviathan's limbs,

   its strength and its graceful form.

13 Who can strip off its outer coat?

   Who can penetrate its double coat of armor [b]?

14 Who dares open the doors of its mouth,

   ringed about with fearsome teeth?...

18 Its snorting throws out flashes of light;

   its eyes are like the rays of dawn.

19 Flames stream from its mouth;

   sparks of fire shoot out.

20 Smoke pours from its nostrils

   as from a boiling pot over burning reeds.

21 Its breath sets coals ablaze,

and flames dart from its mouth...

25 When it rises up, the mighty are terrified;

they retreat before its thrashing.

26 The sword that reaches it has no effect,

nor does the spear or the dart or the javelin.

27 Iron it treats like straw

and bronze like rotten wood...

33 Nothing on earth is its equal—

a creature without fear.

34 It looks down on all that are haughty;

it is king over all that are proud."

Snakes

There was an ancient Egyptian deity named Apep, which was the Serpent of the Nile. This enormous snake was also known as the god of chaos and one of the enemies of the light and truth. Apep is believed to lie just below the horizon line, waiting for a chance to swallow the sun and plunge the world into perpetual darkness. In other Egyptian legends, Apep would sometimes descend into the underworld so he could devour the souls of the damned.

Although Apep is a far cry from the Druks and Wyverns that could be seen in the Eastern and Western cultures, he still points to the perennial fear of snakes by humans.

According to the anthropologist David E. Jones in his book, An Instinct for Dragons, monkeys, dogs and humans, by their nature, are afraid of serpents and other large predators. Because of that embedded ancient fear, people from different parts of the world could have independently concocted their own versions of dragons.

Dragons in Popular Media

"Never give up, and good luck will find you."

- Falkor – The Neverending Story

Dragons have persisted in human history, so much so that they are not just relegated to being in ancient folk stories. Dragons can now be found in books, movies, video games, and many other forms of media. In fact, your fascination with dragons most likely came from seeing one of them in movies, or you read about the magical creatures in a fantasy novel.

If you are wondering at just how much dragons have piqued the interest of modern human society, here is a short compilation of the most famous dragons of recent years that have appeared in various forms of media.

Dragons in Film and Television

Falkor (The Neverending Story)

Falkor is the "luckdragon" that accompanied Atreyu and Bastian in the film series. Falkor is the only luckdragon to appear in the films, but it was said that there were five other in existence in their world. The original name of Falkor in the German novels was Fuchur, which was derived from the Japanese term "Fukuryu", which means "lucky dragon". In the English translation of the book, the name Fuchur was changed into Falkor as it sounds like a certain curse word.

Falkor has a long serpentine body, rudimentary paws, and pink scales and fur, which appear white in dull lighting conditions. The appearance of his head and face differs from the German books and their English versions. In the German illustrations, Falkor would either resemble an Oriental dragon, or he could have a head that looks like a dog (a Great Pyreness to be more exact). On the other hand, on the cover of the English translation of the book, the cover artist envisioned Falkor to look more like a dragon than a dog. In the movies, Falkor had dog-like features and behavior (he like being scratched behind his ears).

Falkor does not have immense strength, nor does he have boundless magical powers (he can breathe blue flames though), but what sets luckdragons like him apart from the others is that they are incredibly lucky. For instance, he was able to locate Atreyu inside a destructive and blinding storm, something that would be quite impossible in normal circumstances. Luckdragons draw air and heat into their bodies through their pink scales, which means that it is impossible for them to submerge themselves in water as they would quickly drown.

Smaug (The Hobbit)

Smaug is the main villain in J.R.R. Tolkien's 1937 novel The Hobbit, which was the precursor to the Lord of the Rings trilogy of books. Smaug was a powerful and vicious dragon that laid waste to the Dwarven kingdom of Erebor 150 years before the events mentioned in the book. In an attempt to regain the kingdom and avenge their fallen brethren, thirteen dwarves, led by Thorin Oakenshield, and aided by the wizard Gandalf the Grey and Bilbo Baggins, a hobbit that was unintentionally dragged into the mess.

In the novel, Smaug is described as "a most especially greedy, strong, and wicked wyrm."

Smaug's origins are a bit sketchy, most of what is written about him only states that he was a fire drake that was born of the Third Age, and that he was considered to be the last great dragon to have ever existed in Middle-earth. Just like the other dragons that came before him, Smaug desired wealth, and he would do anything to get it. The great dragon was instinctively drawn towards the massive coffers of wealth that the Dwarves of the Lonely Mountain accumulated under the reign of the great King Thror. Smaug first laid waste to the city of Dale before heading towards the Lonely Mountain and claiming it, and as a result, he drove all of the surviving Dwarves to live in exile.

In the animated movie adaptation of The Hobbit, Smaug looked nothing like he was described in the books. In the animated film, the great Smaug had very noticeable feline features, including whiskers. On the other hand, in the Hobbit film trilogy, Smaug is not your typical Western dragon, he is more like a wyvern than a dragon, with two wings, two legs, and a serpentine body. This look is more in line as to what Tolkien originally had in mind Smaug would look like.

Drogon, Rhaegal, and Viserion (Game of Thrones)

Drogon, Rhaegal and Viserion are the three dragons that were born in the middle of the Dothraki desert, Daenerys Targaryen. Drogon, was named after Khal Drogo, Daenerys's deceased warlord husband. Rhaegal is named after Rhaegar Targaryen, the last Targaryen to sit on the Iron Throne (he was rumored to have been assassinated and left without an heir, supposedly). Viserion is named after Viserys Targaryen, Daenerys's brother.

Drogon has pitch black skin with red markings. Rhaegal had emerald green skin with bronze markings. And Viserion had cream-colored skin with gold markings. Although Drogon managed to get quite big in the television series, the largest dragon to have ever lived was the gigantic dragon called Balerion the Black Dread, which was the dragon mount of the Aegon Targaryen, also known as Aegon the Conqueror. Balerion was so huge that his skull was bigger than a horse-drawn carriage.

In the television show and in the books, Daenerys came upon the dragon eggs when it was gifted to her by a traveling merchant. At first, the people, including the merchant thought that the eggs were too old and ancient to be able to hatch. However, when the last remnants of the Dothraki tried to kill her by burning her alive in her hut, the three dragon eggs managed to break through their shell, and out of the ashes out came Daenerys, with nary a scratch or burn on her.

Unfortunately, Viserion got killed when the Night King threw a spear that went through his heart. The Night King then proceeded to resurrect Viserion, thus turning him into a dragon wight, one that breathes blue flames. These blue flames of Viserion brought down The Wall and thus enabled the army of White Walkers to descend down south.

The dragons of Game of Thrones are technically not dragons if you are going to get nitpicky. Because the three have two wings and two hind legs and long serpentine bodies, they are actually wyverns.

Shenron (Dragonball franchise)

Shenron is the mystical dragon god that can grant those who summon him three wishes. However, to summon him, the seven mystical Dragonballs must be collected and brought together. After fulfilling the summoner's wishes, Shenron will disappear, and the Dragonballs would once again scatter all over the globe and turn into stone for exactly one year.

Shenron is an Eastern dragon. He has a very long, green, serpent-like body, four legs and four claws on each foot. He has a reptilian head, deer-like horns, and long whiskers.

Shenron can apparently grant almost any wish, as long as it is relegated in just one universe (there are more than a dozen universes in the Dragonball series right now), including resurrecting thousands of lives, and undoing catastrophic damage.

King Ghidorah

King Ghidorah is one, if not the top arch-nemesis of the King of All Monsters Godzilla. Ghidorah is quite unlike most of the dragons in this list in terms of looks, which has not changed much since the first time he came bursting out of Japanese TV screens in the 1960s, he is still an armless, two-legged, golden-scaled, three-headed, two-tailed dragon with gigantic bat-like wings. His origins have varied with his every appearance on screen, from being a planet-killing dragon from another galaxy, a genetic experiment gone wrong from the future, to being an ancient Japanese guardian deity that was awakened by the sudden appearance of Godzilla or the encroachment of humans into the forests.

Aside from flight and the ability to fire "gravity beams" from each of its three mouths, King Ghidora's abilities seem to change with every iteration of the character. In one movie, Ghidora has the ability to shoot lightning bolts from its wings, in another, the monster can quickly regenerate chopped off body parts.

Alduin and Paarthunax (Elder Scrolls V – Skyrim)

Although the entire game centers around the main character slaying dragons, in Skyrim there are two dragons that are most notable. The first is Alduin, which is the malevolent one of the two. Alduin (aka the World Eater) was prophesied to return to Tamriel after a certain series of tumultuous events happened so he can destroy the world and rebuild it once again.

Paarthunax is the younger brother and former lieutenant of Alduin the World Eater. The reason that Paarthunax decided to leave his brother's side is because Alduin claimed that he was a god. Paarthunax decided to reside at the highest point on Skyrim, known as the Throat of the World, or High Hrothgar, and became the leader of the

Greybeards, a group of humans who understand and could use "Shouts" or the dragon language.

Alduin is a pitch black dragon with blazing red eyes. Paarthunax, despite being the younger brother of Alduin, is pale grey in color, and he looks really battle-worn. Despite being immortal and basically eternally youthful, Paarthunax's body is riddled with scars, his wings have tears in some places, and missing a couple of teeth.

Both Alduin and Paarthunax are technically wyverns as they both have two wings and two hind legs. However, their wings also have claws at the ends, and they can crawl on the ground using their wings like front legs, which makes them look like the traditional image of dragons.

The Jabberwock (Alice Through the Looking Glass)

The Jabberwock is the mythical dragon creature that is the subject of the mock poem that Alice found in the book Through the Looking Glass. At first, Alice could not ready the text in the poem, but then she remembered that she is in a mirror world, so the text must be reflected on a mirror before she can read it.

The Jabberwock also appears in the Alice in Wonderland live action movie. In the movie, the Jabberwock was the chosen champion of the Red Queen, and Alice is the champion of the White Queen. According to the prophecy, both champions must fight to end the war that is currently raging on in Underland. With the help of the Vorpal Blade, Alice managed to beat the Jabberwock by outsmarting the huge dragon, leading it to a smaller battleground to limit its movements, and then slicing its head clean off its neck.

Saphira (Eragon – The first book in the The Inheritance Cycle Book Series)

The Inheritance Cycle is a four-book young adult book series that tell tales of different dragonriders and their adventures. The first book, Eragon, was actually turned into a feature-length film, but it unfortunately flopped in the box office, so none of the other books got their own film interpretation. Saphira is the dragon of the main character of the first book, and she is by far the most popular among all the dragons in the book series.

Saphira is the dragon that bonded with the dragonrider Eragon from the moment that she hatched. She is one of only two female dragons during the reign of Eragon II. Saphira, as per her namesake, has scales and eyes that are the color of bright blue sapphires, and they are also said to refract light, making her gleam a little when hit by sunlight, which is why the elves also gave her the name Saphira Brightscales.

Although Saphira, by all accounts, was a young dragon, she was wise beyond her years. Aside from being his partner, Saphira also served as advisor to Eragon and helped him make difficult decisions.

These examples of dragons in modern popular culture is testament that the wonder that humans have with dragons will never fade, even after hundreds of years have passed. Although only a handful of people literally do believe still that dragons exist, the symbol of this mythical creature is so ingrained in society, so much that even your grandchildren's grandchildren will surely have tales of dragons to keep them entertained.

Dragon "Sightings" in Modern Times

"Live in the present. Remember the past. And fear not the future, for it doesn't exist and never shall, there is only now."

- Saphira – The Inheritance Cycle

As mentioned earlier, some people still believe that dragons really existed, that they are not just creatures that existed in fairy tales and folk stories. If you think about it, the world is such a huge place, there are still large chunks of the earth that have not been explored yet, and there are new species of animals being discovered almost every day, so there might be some possibility that dragons actually existed on earth; and there might even be some still roaming in some yet to be discovered place on the planet.

Some people claim to have seen real live dragons, and some even supposedly have "evidence". Over the years, there are have been many eyewitness accounts from people who claimed they have seen real live dragons.

Here are some of the many modern tales of dragon sightings from all over the world.

Dragon Skeleton Found in China (October 2017)

All of the residents of a small rural village in China claim to have found the remains of an oriental dragon. There was even an almost thirty second video of the skeleton in question arranged on the ground, with dozens of residents of the village of Zhangjiakou, in China's northern Hebei

Province, surrounding the somewhat complete remains of a supposed dragon.

The skeleton, fully assembled, had a large skull, and a body that is at least 60 feet long. The "dragon" has four limbs and no wings, which is the traditional look of Chinese dragons. Unlike Western dragons, the Chinese dragon has no wings, but they are said to swim effortlessly through the air and in the clouds.

Sadly, the users of the Chinese social media network, Weibo, claim that the skeleton was most probably placed there on purpose, and the bones are not really that of a dragon's. According to the experts, the bones are too fresh to be hundreds of years old, and that they actually came from another animal, most likely those of cows.

Dragon Caught in Video Flying Through the Mountains of Laos and China (October 2016)

A grainy piece of footage started making round in the Chinese social network, Weibo, in the last quarter of the year 2016. The video, which seems to have been taken using a rather old smartphone, shows the mountain range in China that is quite near the country's shared boundary with Laos.

The video shows a creature flying through the air and going behind a mountain range, a creature that looks just like the dragons that you see in the television show Game of Thrones.

Some of the comments left by the Chinese commenters believed that the creature was not a dragon, but a pterosaur, or maybe a surviving descendant of the dinosaur that retained much of its ancestor's features. However, most of the comments do believe that the video caught a real life dragon.

However, not all were convinced of the authenticity of the video. Many of the comments were very skeptical. A lot of people believe that the creature was made using basic CGI and then superimposed onto the video. The reason, they say, that the video was so grainy is so the CGI dragon can be placed into the video without much trouble.

Dragon Sightings in the Carribbean (March 2014)

A couple of local hikers on St. Kitts, an island in the Eastern Caribbean, reported that they have seen dragons flying around and even inside Mt. Liamuiga, a dormant volcano in the middle of the island. The hikers, who were students from the Ross University School of Veterinary Medicine, said that they saw large shadows looming in the fog around the crater of Mt. Liamuiga while they were hiking their.

One of the hikers said that he thought that the shadows were that of a bird, a very large bird; actually the biggest bird that he has ever seen. However, it could not have been a bird because it was effortlessly moving through the fog at the crater, and the heat up there was too intense for any creature to live there.

This is not the first time that people reported seeing something strange in the area around the volcano. In October 2013, tourists reported seeing flames burst out and hearing muffled roars emanating from the caves at the bottom of the volcano.

These dragon sightings, as the locals would tell you, might explain why the fields in St. Kitts always seem to be on fire. There always seem to be scorched fields every time you drive down the highways, and no one knows how the fires got started. The locals are constantly living in fear that their homes would be the ones that get torched next.

Government officials are quick to denounce that the dragon sightings are responsible for the fires, and they refuse to confirm nor deny the existence of the creature. The officials insist that the fires are controlled burns so they could easily clear fields for development. However, the locals find it hard to believe as these so called "controlled burns" always seem to get so close to villages and towns. Many people still believe that the presence of dragons in the island is a plausible explanation for the phenomenon.

Cryptozoologists from all around the world have flocked into St. Kitts in hopes of getting a glimpse of these mythological creatures, or at least some solid evidence that they are there. These experts are looking for tangible and reliable information, like skin sheddings, particularly distinct scorch patterns, and hopefully a big pile of excrement from the creatures. The latter would be the best discovery because you can actually learn a lot about a creature just from its poop.

If in case dragons are discovered in St. Kitts, the government will have no other choice but to construct a special quarantine area around the base of the volcano. This is to protect these creatures from humans, and to protect the townspeople from further dragon-related incidents.

Ropen, Utah Dragon Sighting (June 2017)

A woman and her 12-year-old son decided to camp out and sleep in their backyard in Ropen, Utah. Before going to sleep, the mother and son duo went stargazing a bit. The pair had no idea that they would lay witness to a huge, dragon-like creature fly overhead, and apparently they did not know that the other townsfolk have reported seeing the same creature earlier in the year.

As the mother-son pair tried their best to identify all the constellations they could find. Suddenly, the boy tugged at his mother's shirt and pointed her upwards, and asking her what is that thing flying in the sky. As she turned her head she saw a huge featherless thing fly right over their backyard. The creature was just flying fifty feet over the ground, so the pair could clearly see it from the head to the tail.

According to the witnesses, the creature looked like a featherless bird with a head that looks like a claw hammer. The two also mentioned that the creature was flapping its wings, so the possibility of the creature being just a dressed up drone is ruled out.

Another eyewitness from the same neighborhood said he saw the exact same creature earlier that year. Just like the other eyewitness account, the man saw the strange creature flying overhead at night, but it was flying much higher this time. However, the man noticed that the creature had a glowing outline, making it easily seen at night, this suggests that the "dragon" was also bioluminescent.

The man also saw the "dragon" fly south, or southeast, towards the cliffs in the mountain range that is just a few miles to the east of the town. The man believes that the creature might be living in one of the hundreds of caves in the mountain range, which is a real possibility.

These are only a couple of the hundreds, if not thousands of eyewitness accounts of dragons. Just the sheer number of these stories, makes it seem that the possibility of the existence of dragons quite plausible. However, there is no solid evidence that supports the theory of that dragons once existed and are still existing in this world, but that does not seem to deter fans of this creature of myth and legend.

www.ingramcontent.com/pod-product-compliance
Lightning Source LLC
Chambersburg PA
CBHW030530220526
45463CB00007B/2775